HOME SPACE CREATIVE DESIGN ROUNDUPS ＋

居家空间创意集 欧式简约
BRIEF EUROPEAN

深圳市海阅通文化传播有限公司 主编

中国建筑工业出版社

序 PREFACE

Jane European Style blends Classical European Style and modern life together. It makes residences luxurious and romantic. In recent year, a ever-growing number of people favour this style in China, due to its elaboration, concision, elegance, nature and fashion. Spaciousness is the elementary characteristic of Jane Europe Style in. Spacious space reflects its momentum. The frech window, the shape of the fireplace, decorative plaster lines... are also strking characteristics of Jane European style.

For another, Jane European style largely depends on the combination of various furniture. Such as the consice sofa, the matching of dark and light colors, showing the clearness and nature. Not only does it conform to the aesthetic standard Chinese, but also it reflect the European style.

This book shows the characteristics of Jane Europe style from various aspects with several highly representative works. Designers are so ingenious that they satisfy Chinese people's taste perfectly and embody the modern fashion.

简约欧式风格融合了古典欧式风格和现代生活的元素，让居室既显得豪华大气，又不失浪漫写意。凭着它精雕细琢、简练优雅同时贴近自然、携手潮流的特质，近年来受到越来越多国人的青睐。简约欧式风格的空间特色首先体现在宽敞、充足的空间足以体现它的风格气势。落地的玻璃窗，壁炉的造型，带有花纹的石膏线勾边，等等，均是简约欧式风格的亮眼特点。

其次，简约欧式风格很大程度上取决于各种家具的组合。如简约的沙发，或浅色或深浅的搭配，清新自然，既符合国人内敛的审美观念，又兼带欧洲风格。西方风情造型的壁灯，金属框裱起的抽象画或是一幅摄影作品点缀在墙上就能营造出浓郁而具有人文风情的欧式情调。

本书精选了多个极具代表性的作品，多方面展示了简约欧式风格的特点。设计师们在此风格上的匠心独运，既满足国人的品位，又展示了现代流行。

居家空间创意集
HOME SPACE CREATIVE DESIGN ROUNDUPS

04 **A**Villa in a Country Region in North of Israel
以色列北部的乡村别墅

14 **A**yazpasa House
Ayazpasa 大宅

20 **N**ove II Villa
NoveII 别墅

26 **P**acaembu Apartment
帕卡恩布公寓

32 **B**ombay Penthouse
孟买的高级公寓

38 **P**hinney Residence
菲尼现代住宅

44 **Q**uiet Countryside
乡村恬静

50 **S**henzhen Longines Peninsula House 9Th Floor
深圳市浪琴半岛 9 楼

58 **A**nting Villa
安亭别墅

64 **M**ix Style Fit in Fashionable Classical
青田苑空间设计

70 **S**henzhen Longines Peninsula House 7Th Floor
深圳市浪琴半岛 7 楼

76 **E**ast Provence Villa
东方普罗旺斯别墅

84 **E**uropean Neoclassical Style
深圳鼎太风华欧式新古典

欧式简约 BRIEF EUROPEAN

目录 CONTENTS

A VILLA IN A COUNTRY REGION IN NORTH OF ISRAEL
以色列北部的乡村别墅

Designer: Hilit Karsh
设计师: Hilit Karsh

Photography: Itay sikolski
摄影师: Itay sikolski

Plan 平面图

The project includes an interior design of a villa in a country region in north of Israel. Interior design work done by Hilit included close escort of the interior design of the house, starts with the stage of planning the skeleton through the interior design, including flooring, lighting, elements made of plaster, kitchen, bathrooms and bedrooms, down to the final styling.

Embedded in a country house outside the openness draws airy spaces which are expressed in broad, bright and clear yet a "retreat" and an object of longing.

Minimalism calculated visual cleanliness, are expressed in the walls free of decoration, still dotted with patches of color that add interest to the space.

The clean white color was selected for the spacious kitchen which its greatness is in its simplicity combined with complementary colorful accessories.

The final product of all of these is a sophisticated space yet with a minimalism of lines and a lot of interest.

该项目是为以色列北部的一幢乡村别墅作室内设计。Hilit 的这个室内设计从规划室内设计的框架阶段，包括地板、照明、石膏线、厨房、浴室和卧室，到最终的风格的定位。

别墅坐落在风景如画的环境中，宽敞，明亮，清新，是一个理想的休闲隐居住所。

简约主义清新简洁的视觉效果，在去除了浮华装饰的墙面上充分体现。点缀着的彩色元素，增添了空间的趣味性。

宽敞的厨房精选了干净朴素的白色，它的包容性与彩色的配饰完美融合。

最终的作品融合了全部的精心设计，既体现了简约风格，也给空间增添了趣味性，可谓匠心独运。

AYAZPASA HOUSE
Ayazpasa 大宅

Architect: Autoban	**Designer**: Seyhan Ozdemir, Sefer Caglar	**Photographer**: Ali Bekman
设计公司: Autoban	设计师: 塞伊汗·奥兹德米尔, Sefer Caglar	摄影师: 阿里·贝克曼

Ayazpaşa house is back again with a fresh new makeover.

Displaying pieces from Autoban's newest collection including the cosy Nest Chair and a specially enlarged version of Deco Sofa, the house has been given a sleeker, more dramatic look. The house has an extremely high ceiling, a dramatic feature that originally set the scene for the design of the space. The kitchen has an ornate gold ceiling as its central feature, with the designs at floor level simpler in form– all the appliances are integrated in a single large marble block. Wood is used for the cupboards for a stunning contrast of materials, while providing a breath of warmth to the space. The marble bathroom bears a strong reference to Hamam culture, and features a massive washbasin with brass legs and a thick marble framed mirror hanging above.

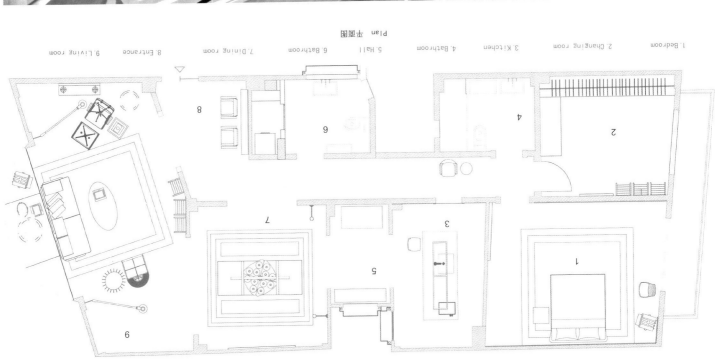

平面图 Plan

1.Bedroom 2.Changing room 3.Kitchen 4.Bathroom 5.Hall 6.Bathroom 7.Dining room 8.Entrance 9.Living room

Brief European 欧式风韵

这套位于伊斯坦布尔 Ayazpaşa 小区以一个色彩明亮雅致的装潢呈现在我们面前。

套房的整体格调来自于 Autoban 的最新系列作品，每处的细节均与其赋有的一种气质形成大胆的对应关系，譬如少许也许上又重新加以注目。的新设计中加进了过多属于当代的一个抽象的绘点，但同时的设计有着保持住的就是所有所有设备的装饰就是在每一再颜色的选择上突出，木材机能之上的层次用过了一种不规则的方式，同时几乎采用以几乎淡来突出气氛。大理石的沙发套和抽屉里存有较多的石材质重要配件，以及几乎形成的几块片石和摆放的格子，体现用几乎都的上与其风味场。

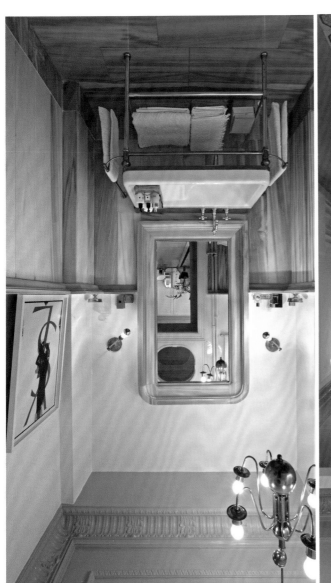

NOVE II VILLA
Nove II 别墅

Design company: Studio B Architects
设计公司：B 建筑工作室

Photography: Raul J. Garcia
摄影师：Raul J. Garcia

This small residual wedge-shaped property sat vacant for years given its many site restraints and complexities including; front streets on two sides, 10'-0" snow storage easements on 2 sides, buried City of Aspen main water and electric lines directly under the building footprint which required re-location and a year of approvals and strict zoning ordinances given its location in Aspen's Historic West End.

The resulting solution is a direct response to these parameters and is a stone clad 'sculpture' which hints to the owners' art collection housed primarily within its double-height space.

The program of 3200 square feet is divided between an upper private level, the public main floor and the lower guest level. The exterior materials consist of a light-weight limestone panel system with aluminum windows and doors.

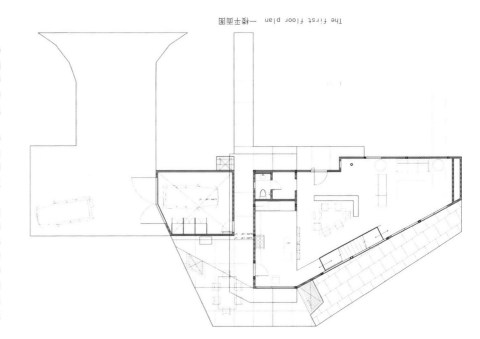

The first floor plan 一楼平面图

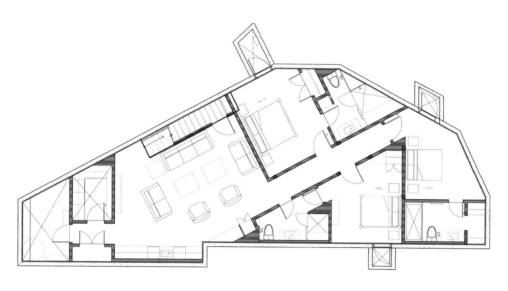

The second floor plan　二楼平面图

这一小块楔形的空地已经闲置多年，由于很多地理位置上的原因，利用这块空地受到很多限制，并且很复杂。包括两边的街道，两边储存室的地役权，以及埋在大楼脚下的Aspen的主要水电线路，这些都需要重新规划，经过一年的审批和严格的区域法令，它们被安置在Aspen城的历史西街区。

解决方案直击这些问题，它是一个被叫做"雕刻"的石头保护层，这就表明了主人把他的艺术收藏室建在了两层高的空间里。

这个3200平方英尺（约297平方米）的设计被分为上层的私人空间、公共区和下面的会客区。外部材料包括轻质石灰岩板和铝合金的门窗。

Attic plan 阁楼平面图

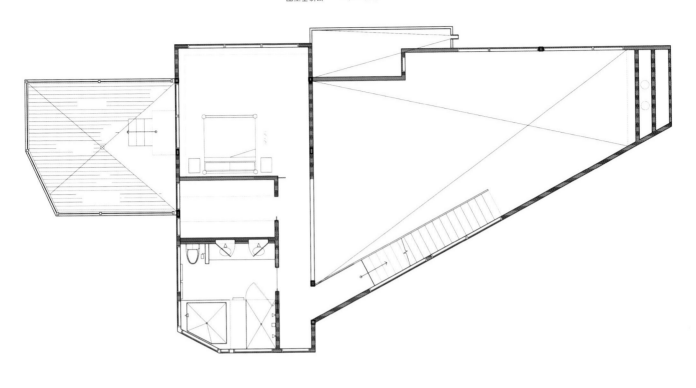

PACAEMBU APARTMENT
帕卡恩布公寓

Design company: Diego Revolloo Architetura   设计公司: Diego Revolloo Architetura

The architecture project for this apartment of 400 square meters, located in the district of Pacaembu, in São Paulo, is characterized by enormous voids that intersperse the remote environments. Before the execution of this project, both the structure and the existing decor had some extremely overshot aspects and could be considered as excessive and heavy. Aiming to promote a feeling of a greater subtlety in the space, firstly, the frames were removed. In addition, new door linings and height skirting boards designed by the office also helped to provide lightness in the social area.

The wide layout was arranged neatly to make even clearer the artwork. Besides the neutral tones on the joinery and tissues were prioritized, took advantage of the quality of sunlight in the apartment. However, in order to break a possible monotony and excess of sobriety, one whole wall was painted black. In it are situated the bookcases and the fireplace (existing), one of the highlights of the living. With it, the ebonized wooden chair (already owned by the clients) makes an interesting entirety.

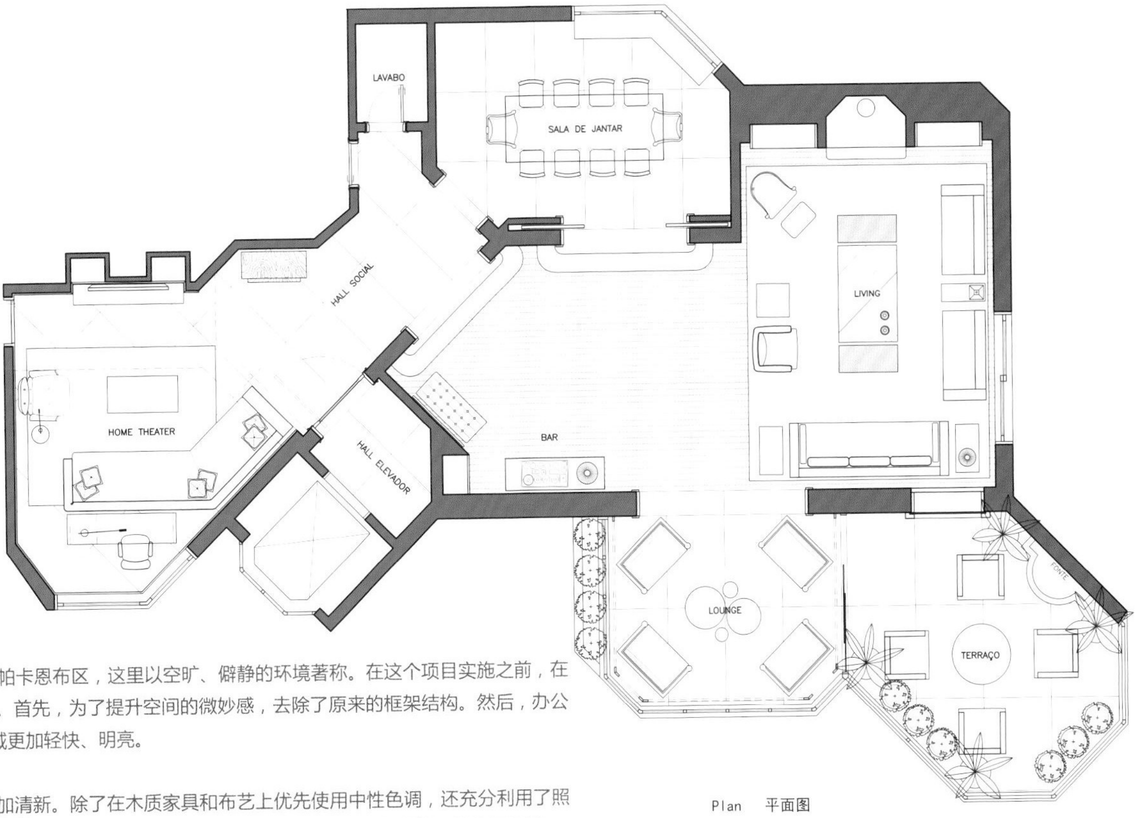

Brief European 欧式简约

这个400平方米的公寓坐落于圣保罗的帕卡恩布区，这里以空旷、僻静的环境著称。在这个项目实施之前，在结构和现有装饰方面有一些多余和繁缛。首先，为了提升空间的微妙感，去除了原来的框架结构。然后，办公区的新门框和踢脚板的设计也使公共区域更加轻快、明亮。

宽敞、整洁的布局安排使得艺术品区更加清新。除了在木质家具和布艺上优先使用中性色调，还充分利用了照进公寓中的阳光。然而，为了打破单调和过度的严肃，整个墙面黑色粉刷。在这里，墙面与书柜、壁炉和谐统一，成为生活中最精彩的部分。一把黑檀木椅（顾客已购置）更使整个环境形成了一个富有情趣的整体。

Plan　平面图

Brief European 欧式简约

The Home Office was determined for the purpose of allowing watch television at any angle of the place. The sofa was moved away from the wall allowing the creation of a new space for work and studies. Its irregular shape demanded a custom designed.

The decor is a mixture of different styles in the right dosage, giving more personality to the project with a mix of contemporary models, parts design and classic style. Some of the furniture already belonged to the family. In the dining room, for example, the original style chairs Jasen were already part of the old layout. With the new wallpaper that imitate linen, the silver velvet in the chairs, the smooth lining and imposing skirting boards created a great contrast with the furniture. The bar table (antique Italian deco commode) also received a new aspect when lacquered in black.

家庭办公区的设计使得在室内的任何一个角度都可以看电视。搬走了靠墙的沙发，为工作和学习创造了一个新的空间。这种不规则的形状需要定制才能完成。

室内的装饰合理地融合了多种风格，给项目带来了兼有时代典范、分区设计和古典风格的个性化特征。一些家具已经成为家庭的一部分。比如，餐厅里自然风格的椅子Jasen就已经成为布局设计中的一部分。仿亚麻的新墙纸，椅子上的银色天鹅绒，丝滑的内衬和富丽堂皇的踢脚板都与家具产生了强烈的对比。漆成黑色的吧台（意大利传统矮柜）同样让人眼前一亮。

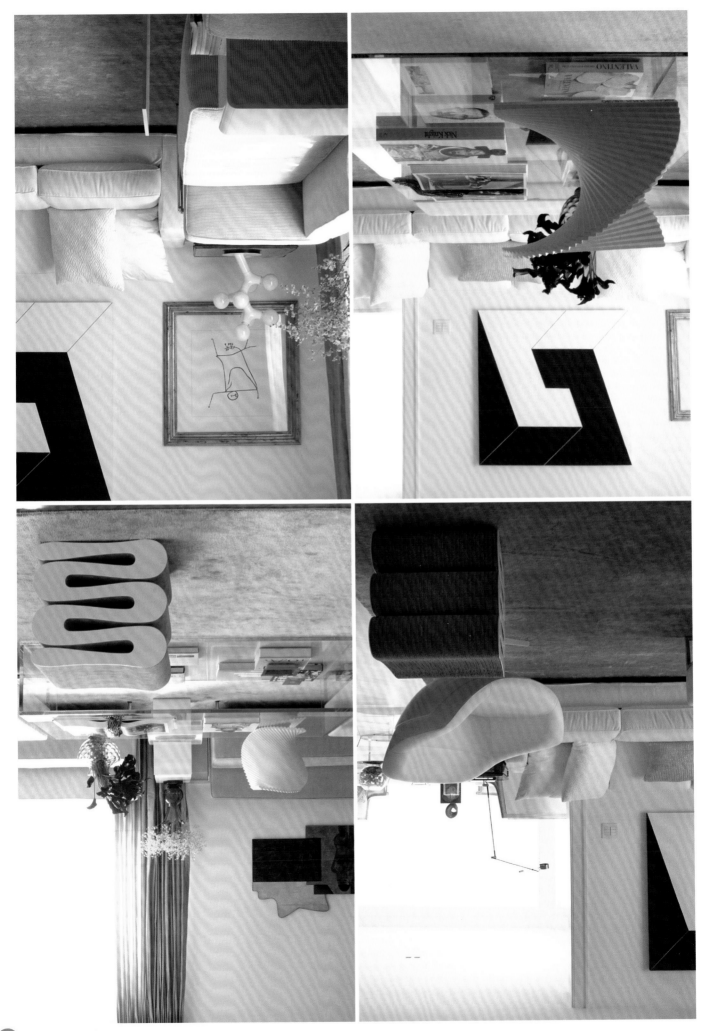

BOMBAY PENTHOUSE
孟买的高级公寓

Design Company: Craig Nealy International Ltd.　　　　　设计公司：克雷格国际有限公司

The first floor plan 一楼平面图

This ten thousand square foot duplex is located in a South Bombay residential tower. It occupies half of the lower footprint on two floors. Craig Nealy International Ltd. designed the architecture of the apartment from the exterior curtain wall in. The team went on to design every piece of furniture, consult with designer Krishna Mehta on all the textiles in use, and curate a photography collection found throughout the residence.

The living room is a two-story volume paved in polished Carrara marble and finished in oyster Venetian plaster. To the left is the open dining room; to the left of the open library. The entire north wall of the living room is a two story glass wall with an operable folding glass wall. This leads to a terrace as large as the living room with views of the Bombay racecourse to the west and of the Arabian Sea to the right.

这套错层有一万平方英尺（约 929 平方米）的双层公寓坐落于孟买南部的住宅楼顶楼，两层共占据了整个楼顶区一半的空间。

Craig Nealy 公司从幕墙之外展开了室内的整体设计开始，设计团队为套房内所有的家具，设计师 Krishna Mehta 为套套房内所有的织物设计，购置散布于整套房之中的摄影艺术品。

客厅有上下两层，地面铺以抛光的 Carrara 大理石，饰面为用珠光灰色的威尼斯石膏涂料。在左边为开放式的餐厅，再左手边则为开放式的图书区。整个客厅的北墙都是一面双层玻璃墙和一扇可手动开合的玻璃推拉门。这样就使得该区向外是一个和客厅一样大的天台，可以观望到孟买赛马场的西边和阿拉伯海。

The effect is that of being on an airplane.

All of the public spaces are in neutral colors with large format color photography on the walls. The four bedrooms, media room and pool are all done in vivid colors and have black and white photography. The dining room seats twelve and has chair upholstery heavy with gauffre and beaded medallions. The living room is divided into two conversation areas: four massive armchairs for curling up as Indians love to do, to the west, and a sofa and four chairs to the east.The refinement of the room furnishings, the carpets and the curtains play off the visceral, photo essay quality of the photography, which depicts the inhabitants of Bombay in daily life. A vast curtain and black sheer open to reveal the glass wall and the view with electric motors. The kitchen is a mix of indigo lacquer and white Carrara marble. It has a powerful exhaust hood and can be completely insulated to contain the powerful aroma of Indian cuisine. The table has been set for a classic south Indian meal.

The library doubles as a guest room. The large bookcase immediately inside the door swings 90 degrees to conceal the library space. There is a fold down bed and access en suite to a ¾ bath.The ground floor bedroom has a flamboyant floral-paisley carpet and a glamorous dressing table.

The gray bathroom is all thermal granite. Its freestanding tub sits in a bed of river rocks. An eggshell marquetry bowl holds bath towels. The master bedroom on the upper level has a mixture of large paisley figures and medallions in pale blue. There is a vanity flanked with glass floor lamps and a chaise longue. Flamboyant curved pulls give access to the wardrobe. The master bath is Nero marquino marble with a bathtub nine feet long flanked with vessels of white marble from Rajasthan.

所有的文字说明均用了中外各国并配上精巧别致的图案的色图片，整体看起来和谐统一，色彩斑斓，别具一格。家中的照片大部分都来自主人的各国旅游留念。为了增加墙体的装饰效果和欣赏情趣，将每张十二英寸的相片，每个装饰相框都精心设计的图案装饰，有时甚至一个精美的架子，有大小不同的挂件和摆件区，增加了视觉感官上的观赏性，吸引了众多前来观赏的客人及参观人员纷纷驻足欣赏。走廊中全都铺设的是地毯，可以为居住者消声，同时也使居室显得典雅舒适，紫色的薰衣草与绿色的莹莹草;

1. 目及至了人为精排列区域，包括了几个大小不等的小巨人的分巨大、使门口外部和内部墙上的门小顶灯箱均为图书杂志作为艺术装饰用，巨人的书架高达90。便可以方便拿取和放置到门口顶部嵌海图书

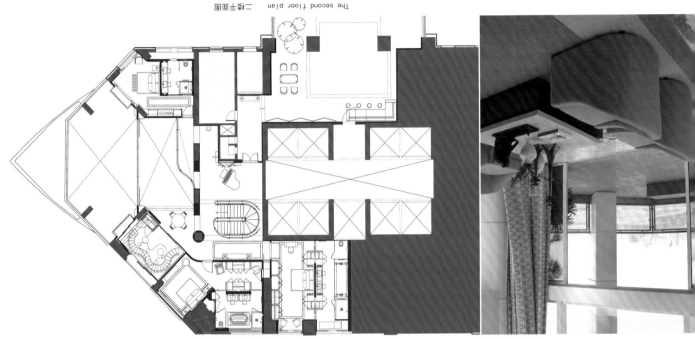

二楼平面图 The second floor plan

的客厅，一楼整体主要以设置做为家中的分区之一，一楼的客厅吸引来众多品茶的朋友在上面。海方交谈的大沙发的沙发上，精致的重厚装饰后的空间家居显示入客的高贵方。

这里的客厅是以家居陈设为主片，海方交谈和方子的沙发和地设计的，精致的重新图案的家具摆设图案的图画上。海方交谈的大沙发的沙发上，精致的重厚装饰后的空间家居显示入客的高贵方。大声，将海里地毯引起了来人出身的人重量方向，大小长的沙发印地脚边摆放了精巧且时尚的各种大摆件。

PHINNEY RESIDENCE
菲尼现代住宅

Design company: Elemental Design, LLC
设计公司: Elemental Design, LLC

Area: 2510 sq ft (living space)
面积: 2510 平方英尺（约 236 平方米）（住宅面积）

Location: Seattle, WA USA
项目地点: 美国华盛顿州西雅图

A young couple with a growing family searched for months to find a site to build their new home on. After many months and numerous near misses, they abandoned conventional channels and approached a homeowner with an oversized lot in the Phinney Ridge neighborhood of Seattle. A portion of that yard was a separate legal lot, left over from an earlier short plat that became the site for their new 2510 SF home.

The design necessarily began by addressing the challenges of a small lot, with significant topography and a large easement that maintained an existing driveway. The resulting footprint dictated that a satisfactory living level could only be accomplished on the third floor by means of a 16' cantilever. This design challenge was to maintain a connection between the street and the living level through the entry while also maintaining the privacy of the sleeping rooms. This was accomplished by carrying exterior materials through the circulation spaces and providing visual clues to circulation at the entry by placing glass under the stairs. With limited yard, locating usable outdoor space proximate to living areas was also especially important. A series of interconnected decks flow from the living and kitchen spaces, leading to the more expansive roof deck. The main volume of the home rises about the street and makes a gesture inward. The subtle shift of the volume turns to access the distant views, provides a buffer from the nearby intersection, and gives spatial definition to the open planned living spaces.

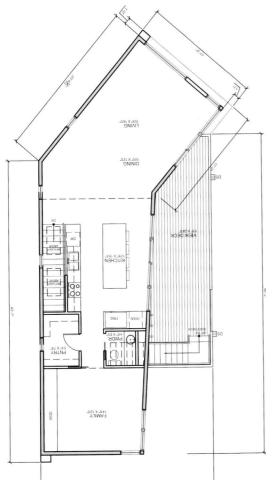

The first floor plan 一楼平面图

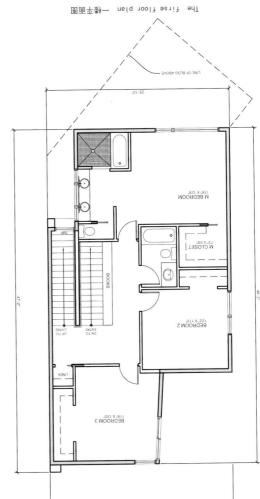

The second floor plan 二楼平面图

一对年轻夫妇带着他们日益壮大的家庭为了寻找一处合适的空地来建造他们的新家已经用了数月时间。几经周折，他们终于抛开传统方式，在西雅图的菲尼山脊上找到一个拥有多余空地的房主。除了独立的部分，庭院剩下的部分就由开始的高地变成建造他们2510平方英尺（约233平方米）家的地方。

这个设计一开始就面临很大挑战，在如此复杂地形结构上保留已有的曲径车道，也就是说符合舒适要求的居住层只能通过一个16英尺的悬臂结构在三楼完成。这个设计挑战在于保证街道和居住层入口的连通，同时还要保证卧室的私密性。这是通过从交通空间搬运进来的外部材料和为交通空间的入口处提供视觉线索，再把玻璃放在楼梯下面得以完成。这样的设计使得住宅能利用大部庭院，占有和居住面积近似的室外空间也同样重要。一系列相互连通的平台从起居室和厨房空间展开，提供了更加宽广的屋顶平台。这种巧妙的空间转换，给交汇空间提供了一个缓冲区，给开放空间带来了新的定义和思路。

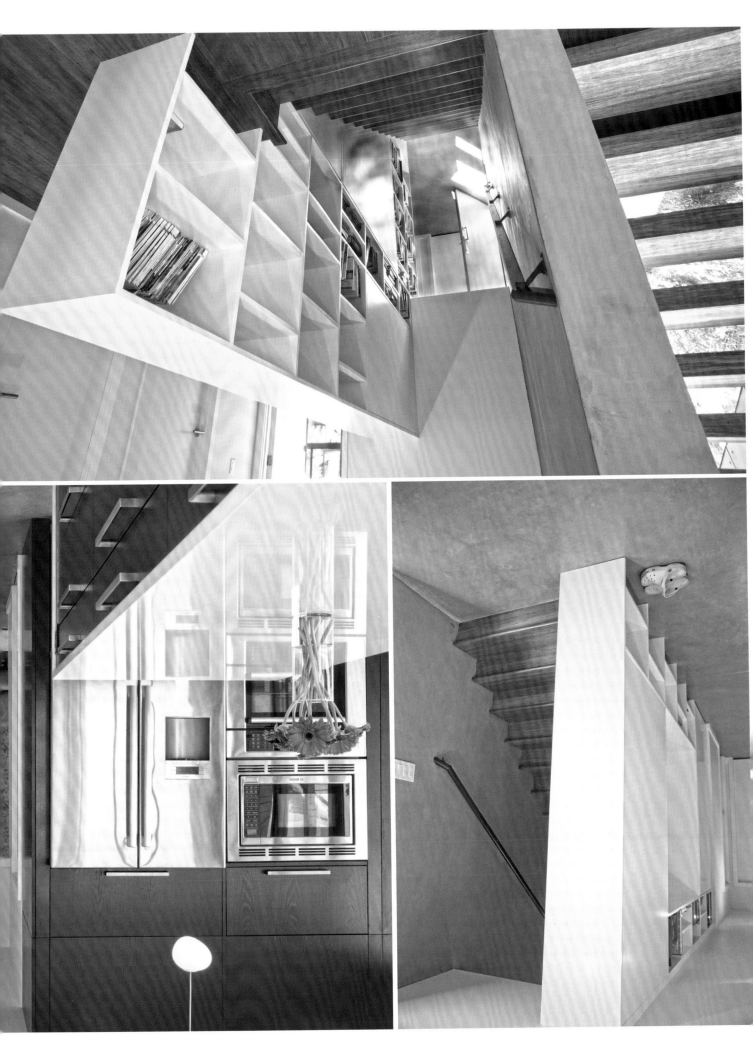

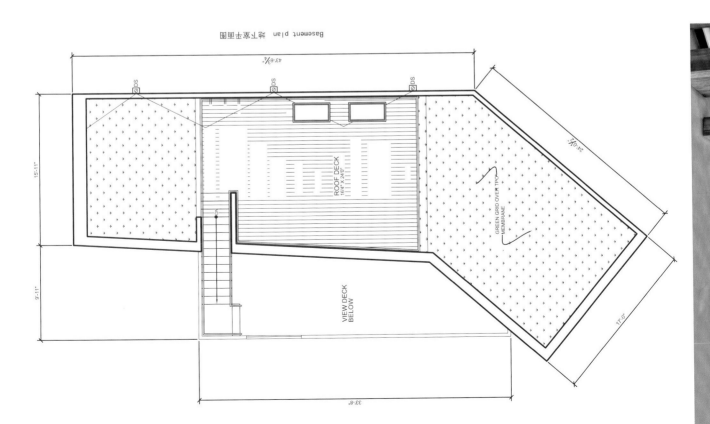

Basement plan 地下室平面图

QUIET COUNTRYSIDE
乡村恬静

Designers: Huang Xi
设计师：黄希

Location: Kunming Yunnan
地点：云南昆明

Area: 120 m²
面积：120 平方米

The main tone of this decoration is white, which feels simple, clear and soft. At the same time the white color represents order and tidiness as well as the unique style of the owner, yet it is not the color which makes you feel dull and boring. The newly aristocratic wallpaper and the decorative lighting in the sitting-room create a gentle and noble atmosphere, meanwhile the furniture, accessories etc. bring tranquillization and nature of countryside into it, making the feeling of good quality and leisurely life. There is full of poetics when the following objects gets together: soft and curled sofas with the pattens of little follower and line, europeanized cushions, a classic tea table, a nostalgic and bright carpet, crystal and gorgeous chandeliers and the warm green. It is the miniature of Europeanization in rural areas and the civilization in the classicality.

Home Space Creative Design Roundups 居家空间创意集

Brief European 欧式简约 47

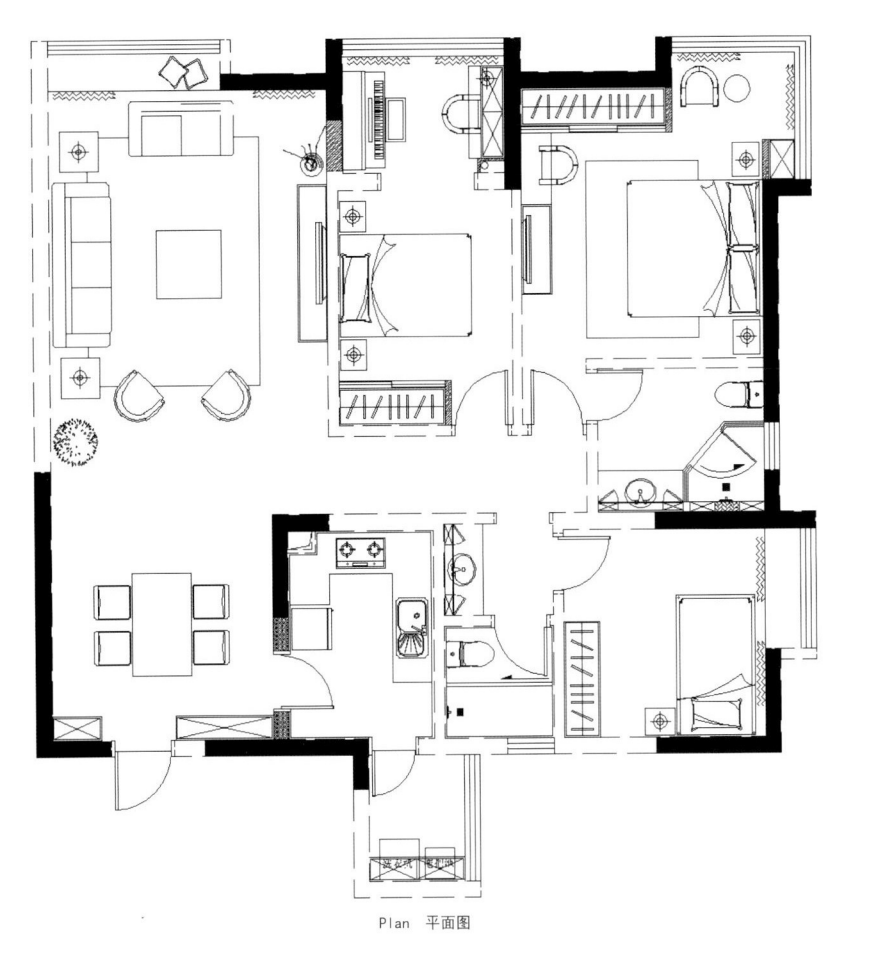

Plan 平面图

整体以白色为主调装修色调，给人简洁，爽朗，轻松的感觉。而一色的基调同时又体现出秩序与井然，彰显出主人独有的风格，但这些绝对不会让你觉得单调无趣。客厅中极具新贵气质的墙纸与灯饰把空间的基调体现得温婉高贵，而设计用家具、配饰等将乡村风格的恬静与自然融入其中，既不失品质又有生活情调。搭配上柔软卷曲带有小花与线条的沙发，欧式的抱枕，古典的茶几，色彩怀旧而不失明快的地毯，晶莹而华美的吊灯以及那暖心的绿意，一切变得那么饱满而富有诗意。乡村中的小欧式，古典中的现代人文。

SHENZHEN LONGINES
PENINSULA HOUSE 9TH FLOOR
深圳市浪琴半岛9楼

Designers: Zheng Hong Xu Jing
设计师: 郑鸿 徐静

Area: 260 m²
面积: 260平方米

Photography: Xu Jing
摄影师: 徐静

A Private house means an elegant life, this house tend to be a neutral style. The house uses light brown as the elementary tone to highlight mild and implicit character. Living room has an elegant atmosphere and simple colors, a maroon safa is the highlight. Straight lines create a clear space. The golden picture frame, black-and-white sketch and other soft decorations form a harmonious picture. Western kitchen in black and white make a strong modern sense. The novel black glass ceiling with circle patterns, and the table and chairs with radian are all ingenious. The picture of collector's Edition HERMES hanging on the wall shows owner's pursuit of the high quality life.

男孩的卫生间中的风格与女孩的卫生间是一脉相承的，以浅咖啡色为主调，本和色内敛，多了些许的质感，淡淡的色彩，搭配几幅装裱起来的明亮色泡图案的图画，黑白相间的瓷砖，打马赛克强烈的对比，构成一幅有主题又丰富的图案，以窗、自然的为主的墙面，再与玻璃隔离加强辨别度的图形案，加之简易强劲的墙裙以及几何图形的汇聚，给首空色心，走廊的墙上移植绽放者几朵中挂画着园的景象与人物海的笑泛泛泛来。

ANTING VILLA
安亭别墅

Design company: KRT **Location**: Shang hai **Area**: 500m²
设计公司: KRT 项目地点: 上海 面积: 500 平方米

Anting house is a new building in suburb Shanghai.The four-people family want a peace and warm home in this busy city.They want a modern idyllic style house.We get through the space to creat a larger public space and make more sunshine enter the house.All of oak floor ,earth tone and classic furnitures diffuse a warm atmosphere.

安亭别墅是上海郊区新建筑,客户是活泼年轻化的都市一家四口,希望寻找一处平和温馨的地方。他们希望我们的设计既富有时尚气息又好像田园间,把居住的大的公共区扩大,并让更多的阳光进入房间。橡木地板和大地色以及经典的家具勾勒出一种温暖的气息。

MIX STYLE FIT IN FASHIONABLE CLASSICAL
青田苑空间设计

| **Design company**: Qingtian garden space design company
设计公司：Hilit Karsh | **Location**: Xin bei City
项目地点：新北市 | **Area**：75m²
面积：75平方米 |

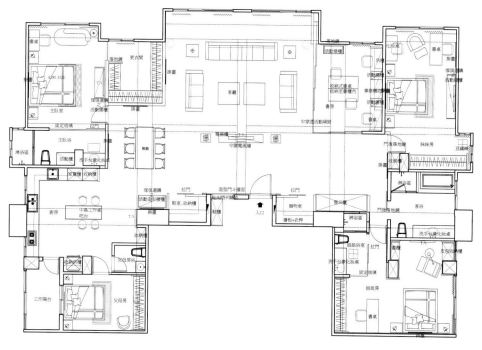

Plan 平面图

The owner have bought two units, and getting through the wall to create a open space so that we can enjoy the scenery and the lighting outside freely. But, the new house has problem in generatrix, so, designer on the premise of guaranteeing.

The designer change the space from lengthwise to crosswise. TV wall was reset between living room and hallway, dining room was moved close to the recessed space of the open kitchen. We can reach living room, study, dining room and bedroom through the enlarged hallway directly. The whole space has a harmonious propotion after redesigned.

Simple lines create a symmetric and full space intention. Custom furniture and gorgeous decoration enhance a taste. They are the meaning of art deco.

Brief European 欧式简约

屋主在购入新屋前，就在预售阶段向建造商表示要将同时购入的两户打通，让屋外绝美的绿意山景及采光，自由流通于规划为开放式的公共空间里。但打通后的狭长新屋有了格局动线不良的情形，在不动主要格局的前提下，设计师将原本客厅靠墙沙发及电视墙位置转向，将纵长的公共空间改为横向发展，并将电视墙移到客厅与门厅中间，将餐厅移到靠近开放式厨房的凹形区块，重新配置过的整体空间比例方正完整，加大的门厅可不经过其他内室，直达客厅、书房、餐厅及卧室，没有一丝矫情多余的空间配置。

以简单的几何线条交会出对称饱满的空间意象，以订做家具及贵气华美的钻饰带出空间品位，是低调奢华的 art deco 表情。

SHENZHEN LONGINES PENINSULA HOUSE 7TH FLOOR
深圳市浪琴半岛 7 楼

Designers: Zheng Hong, Xu Jing
设计师：郑鸿 徐静

Area: 260 m²
面积：260 平方米

Photograph: Xu Jing
摄影师：徐静

Red color may be the most ingenious feature in this house. Red has the special meaning for Chinese,it has been given lots of spiritual meanings,like wealth,happiness,luckiness,etc. Red expresses Chinese culture and guides us to pursue beautiful lives. Maybe because the shareholders' particularity and the Chinese consistent disposition,as consequence, Ouyang family show special preference to red.

As we know,we must be prudent to use red in the space,because too much red likely to cause eyestrain. We consider this issue from many aspects in order to get a suitable choice.It had been thought over and over,drawings had been modified again and again.Finally,we still adopted the original idea.

在这套房子里，最别出心裁之处莫过于对红色的运用。红色，对于中国人来说有着特殊的意义。富贵、美好、吉祥，那一抹红被赋予了许多精神层面的意义。那是流淌在炎黄子孙骨子里的血脉文化，是对生活无限追求的亮丽图腾。或许是作为一个股民的特有讲究，或许是承袭了中华儿女的一贯秉性，总之，红色成为了这一家情有独钟的颜色。

做过设计的人都知道，红色在空间中的运用必须十分谨慎，因为红色过多往往容易造成视觉疲劳。怎样让红色贯穿到整套房子而又不为过，设计师作出了多方面的考虑。设计图纸几经修改，又几经反复，斟酌权衡下还是采用了最初的方案。

EAST PROVENCE VILLA
东方普罗旺斯别墅

Design company：Rui Zhi Hui Design Company	**Designer**：Jiung-Chin Wang；Peng Qing	**Location**：Beijing
设计公司：睿智汇设计公司	设计师：王俊钦　彭晴	项目地点：北京

Dancing art villa——touch your soul

The original inspiration comes from free curves of Guggenheim Museum.These curves are full of imagination.They build an environment which is full of life and light and stimulate people's dream and happiness.The villa is divided to three spaces, spiritual space is fluency like a stream,high space is powerful, visual space has the sense of layering.The mild environment and contrast color are in harmony.With these selective materials,like...we create an unparalleled art effect of space.

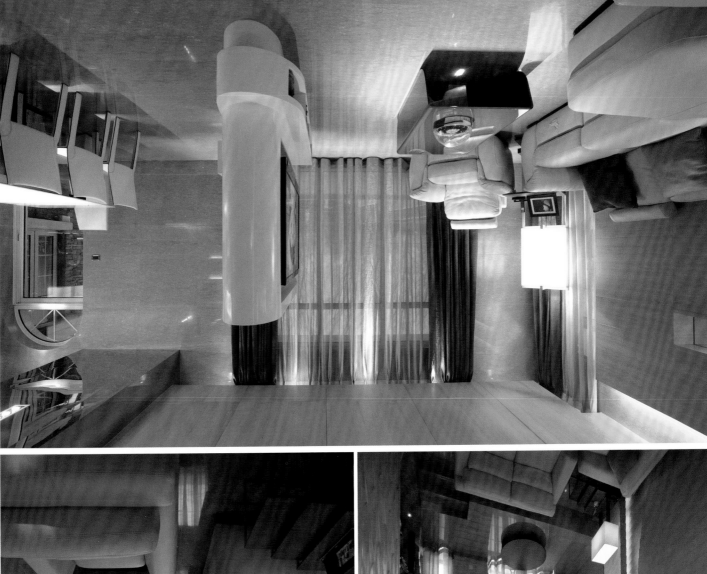

The first floor plan 一楼平面图

舞动中的艺术豪宅——触动心灵的设计

项目最初的灵感来自古根海姆博物馆的自由曲线的元素，线条丰富且充满想象，能够营造出一个充满生活化与光线的环境，激发人的梦想和幸福。整个别墅被划分成三个空间层次，通过区分各自的精神空间、用整体流畅的动态曲线，就像细水流线样贯穿始终、挑高空间的震撼力、视觉空间的层次感，柔和与对比色系的协调，饱和丰富来展现，搭配精心挑选过的材料：千丝万缕石材、千层玉、橡木饰面板、黑檀木饰面板、银箔、马鞍皮革、夹丝玻璃、镜面不锈钢、贝壳马赛克等冷材质为主要材料，这些材料设计出了一种无与伦比的整体空间艺术效果。

Home Space Creative Design Roundups 居家空间创意集

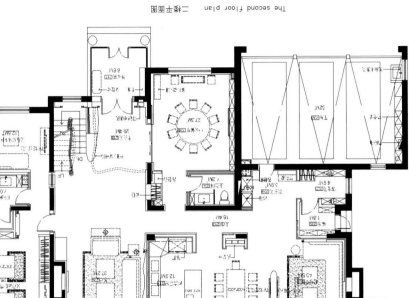

The second floor plan 二楼平面图

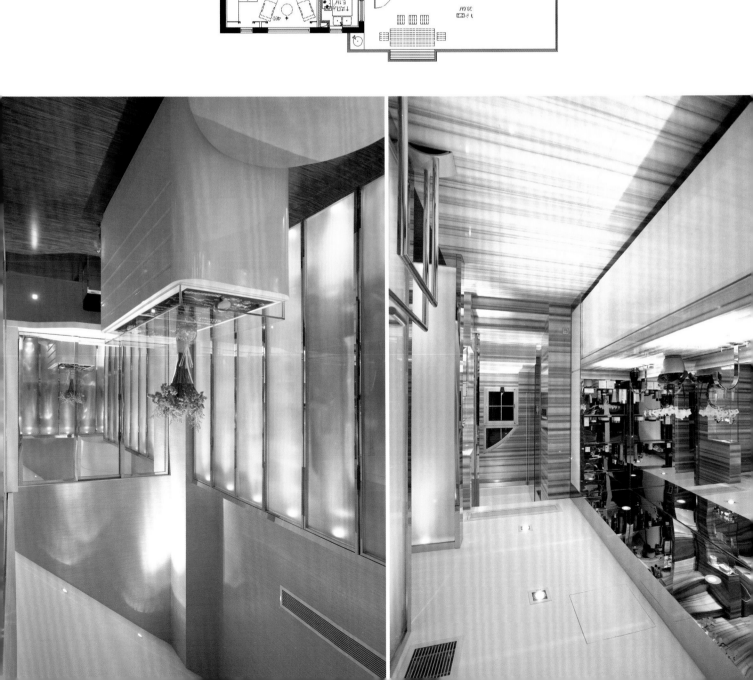

EUROPEAN NEOCLASSICAL STYLE
深圳鼎太风华欧式新古典

Design company: Hong Kong Bai Li International Design co., LTD
设计公司：香港百利国际设计有限公司

Designer: Xiao Qiang
设计师：肖强

This is a double-family house, the former spatial structure was tight and disorganized. Under the premise of basic function, designer remodeled the structure completely and created some new accessary spaces. Such as a independent hallway, a cozy bar, a personalized toy room. The reasonable layout make two unit mix together perfectly. Fuctions have a good relationship with space, ervey room become more spacious and airier. Furnitures are exaggerated but elegant which emphasize the high quality. The whole space shows a balanced visual quality in a light color environment.

本案为双拼结构，原空间紧凑且杂乱无章。设计师将空间结构进行了大刀阔斧的重组和改造，在满足家居基本功能的前提下创造出了原本并不具备的附属空间：独立而大气的门厅、舒适惬意的酒吧区、个性十足的玩具房。合理的空间布局让原本两套不相干的小单元完美融合，同时充分发挥功能空间之间的关联性，将每个空间处理得更宽敞，流线更顺畅；整体环境以浅色调为主，适当搭配造型夸张而不失典雅的活动家私，着重考究家私的质感和品质，如此达到良好的视觉比重平衡，不偏不倚，恰到好处！

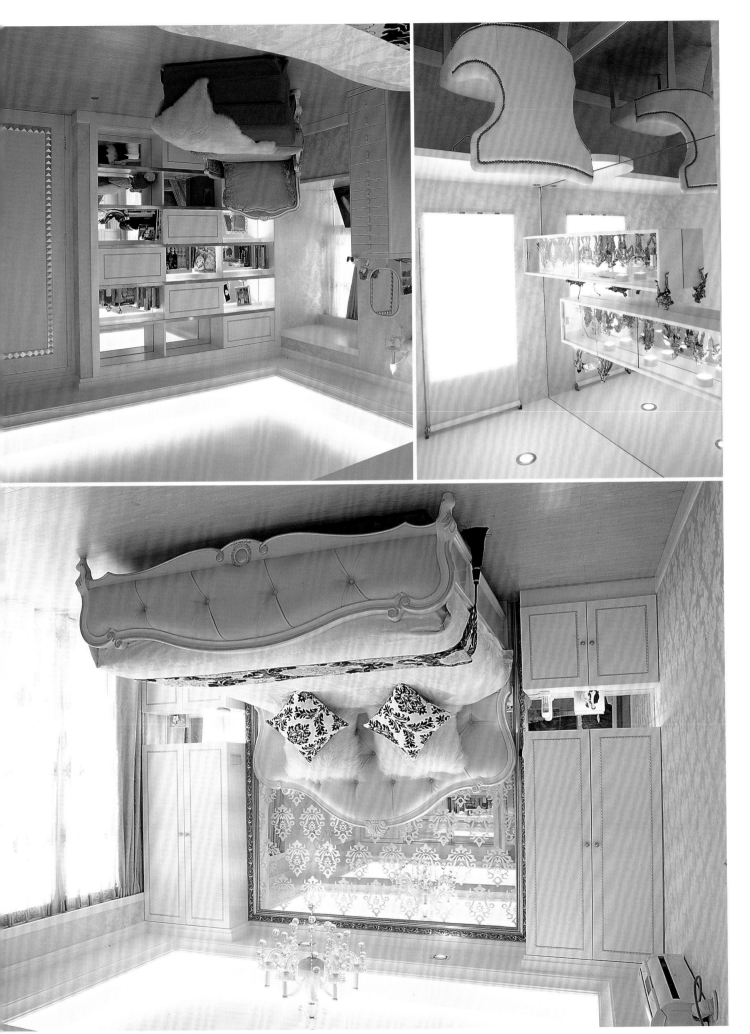

图书在版编目（CIP）数据

欧式简约 / 深圳市海阅通文化传播有限公司主编.
北京：中国建筑工业出版社，2013.4
（居家空间创意集）
ISBN 978-7-112-15182-0

Ⅰ.①欧… Ⅱ.①深… Ⅲ.①住宅—室内装饰设计—图集 Ⅳ.
①TU241

中国版本图书馆CIP数据核字(2013)第038774号

责任编辑：费海玲　张幼平　王雁宾
责任校对：姜小莲　陈晶晶
装帧设计：熊黎明
采　　编：李箫悦　罗　芳

居家空间创意集
欧式简约
深圳市海阅通文化传播有限公司　主编
*
中国建筑工业出版社出版、发行（北京西郊百万庄）
各地新华书店、建筑书店经销
深圳市海阅通文化传播有限公司制版
北京方嘉彩色印刷有限责任公司印刷
*
开本：880×1230 毫米　1/16　印张：$5^1/_2$　字数：180 千字
2013 年 6 月第一版　2013 年 6 月第一次印刷
定价：29.00 元
ISBN 978-7-112-15182-0
　　　（23276）

版权所有　翻印必究
如有印装质量问题，可寄本社退换
（邮政编码 100037）